"You Have Got To Know .... Mathematics"

A reflection as to why mathematics is so important in our lives

---

—Dhaval Vyas

PARTRIDGE
A Penguin Random House Company

To order additional copies of this book, contact
Partridge India
000 800 10062 62
www.partridgepublishing.com/india
orders.india@partridgepublishing.com

Dedications

I dedicate my first book to my family, teachers, friends, colleagues and all those people who have directly or indirectly influenced me. Above all, I thank god for being kind and ever merciful.

I would also like to thank all those whose blessings have helped me reach where I am now.

"Success is not what *I have* and *you don't*.
It's about what we *share* when we never had *any*
and also when we have *some*.
It's a *journey* where we find our *true self*
and have *fun* along the way."
-Dhaval Vyas

You Have Got To Know . . . . Mathematics

CONTENTS

Acknowledgements

I have already dedicated this book to all those who matter in my life. There has been obvious mention of family, teachers, friends, colleagues and all those people who have directly or indirectly influenced my life. I do acknowledge that all through the years in school, my English teacher taught me the basics of English grammar. My Principal, Mrs. Alice Joseph, a stern disciplinarian was a very sweet lady and instilled in us the basic understanding of good English. I cannot forget my mathematics teacher Mrs Shankari for instilling in us much needed confidence.

There were many instances when I felt the need to question my teachers and there have been patient replies much to the annoyance of my friends. In fact, my curiosity was sometimes too much. Nevertheless, they have always put up with my umpteen questions. I acknowledge the encouragement and faith shown by my parents who always thought I was special, though I was perhaps different and difficult. I remember the times when I used go through the academic books for the underlying meanings of the links which I was very sure were missing.

I would like to thank everyone for making this book possible and apologise if I have forgotten to mention anyone who has added to my knowledge.

At the end I wish all the very best to the readers in their pursuit of understanding this very beautiful subject called mathematics.

DHAVAL VYAS

Preface

I have always been fascinated by mathematics as a subject. I named this book "You Got To Know Mathematics" due to its immense importance in our daily lives. In today's world we cannot imagine knowing little mathematics or rather not using it in our lives. Sometimes we use it even without knowing it has been used. What I want to stress upon is the aura and awe that surrounds mathematics. What prompts us to learn this subject in school, college and university? Why do parents and educators lose sleep over inculcating a love for mathematics in their children and students? What happens when one doesn't know mathematics? What is the teen and peer pressure associated with it? How does it affect employers and job seekers? What is its effect on business due to mathematics? What is in it for anyone who comes in touch with mathematics? What are the solutions and how should it be taught? How should the process of learning be initiated? What is the future associated with it? Why do some excel and some fail in learning mathematics?

These are some of the questions which I have tried to address in this book. It is my journey into the very basics of learning the subject which I am sure many of our young readers and also the older ones have also faced when in their youth. You may find a lot of questions being asked and answered throughout the book. They may be relevant to you and your life or simply an overview to give you a slight insight into the workings behind the process of teaching, learning and understanding maths. I will include a lot of philosophical analysis which may seem clear but it is this very complexity associated with maths which has created an aura around it.

Though I am not an educationist and an expert on the subject it is but a journey which I want to share with my readers. And my goal is to simplify and lay identify the difficulties associated with it and take a fresh perspective towards the subject for the benefit of the masses.

I hope you enjoy the book.

Introduction

In this chapter I am going to introduce you to the traditional and contemporary approaches to mathematics and what the future can hold for us.

Traditionally, mathematics has been an elite subject which not everybody masters. You can either be born smart or be an outright dumb person. All throughout the book there is this discourse setting the subject free from the clutches of its traditional approaches, the difficulties and opportunities. There are some serious thoughts to ponder upon throughout the book and it's open to the reader to analyse and decipher and interact with the thought process to arrive at their own conclusion.

We have at one hand the traditionalists who are very adamant about their approach towards the subject and then there is the contemporary approach which is ready to continue serious traditional approach minus the stereotypes and also include some new experimentation in teaching, learning and understanding the subject. I am going to introduce to you to this wonderful subject with some thought-provoking questions and their answers. You may agree or disagree but it will certainly make you more aware of the working behind the subjects.

All throughout the discourse, the problems and opportunities is subtly disclosed so as to give one an insight as to how to face a given situation. I will first take up the history of mathematics and then we will deal with contemporary issues through a series of question and answer series and then we can discuss the future through the last few chapters.

Mathematics has been the quintessential struggle of the rational mind to turn the abstract into a concrete and tangible fruit of its labour. Here we are going into the history and then trying to slowly evolve into something concrete and more current in its approach which affects everyone directly.

I will elaborate more in the next chapter which discusses the history of mathematics.

History—I

In this book I am going to use as simple language as possible so that everyone can comprehend this much misunderstood and feared subject. However, we have to approach it without any prejudice to get maximum benefit. How do you distinguish between numbers and alphabets? Why are numbers different from texts or alphabets? Well, the basic difference is that letters have more to do with associating things by giving these things names. A language is created by using various sounds and asphyxiations. This allows us to speak and get a mental picture of every different thing. Now numbers are also similar but they have more to do with calculations and attachments. We do not know whether numbers were invented first or languages came first. Both could have evolved simultaneously as humans used their surroundings to their advantage.

We can say mathematics evolved in various places and ages such as Mesopotamian civilization, Mayan civilization, Aztec civilization, several African civilizations, Egyptian civilization, Greek civilization, Chinese civilization, Indian civilization, Islamic civilization, Roman civilization, in medieval Europe during the Renaissance right up to the modern day. In fact, mathematics perhaps first evolved when mankind was part of the primate family.

I am going to discuss the contribution of each civilization and period in brief. There is no order and the list does not cover all the branches and discoveries but it gives us some idea as to why or rather how mathematics is where it is now and where it is headed for. There may or may not be a pattern but the need to categorise the vastness of the subject itself shows how compelling is its girth and scope.

Primitive man: Let's talk about the pre-historic times when human beings were actually primates. Did they use numbers and, therefore, maths? Did they actually calculate? Well, there have been experiments showing that even animals have an understanding about their surroundings and can calculate. They also communicate. In fact, since we cannot understand their language we call it unintelligible. However, it is essential for their survival. Pigeons and some birds and animals such as bats, swans and migratory birds are able to calculate the distance to their destinations wherever they may be by using the earth's magnetic field. They are known to use the same pattern and path every time year after year to reach the same spot. Now, this involves calculation though one can dispute this by saying that this capacity is in-built. Apes and chimpanzees have been known to make tools to fetch their food. Crows are able to use tools to get their food. Ants, honey bees are still smaller living things but are known to follow a path to reach their ant-hills and beehives. All this indicates that all living things have an intelligence which may be primitive as per our standards but which help them to derive many benefits.

This brings us to the question: Is calculation conscious or abstract? Well, as we understand, our brains have both abstract functioning and conscious decision-making capacities. So do the various living beings around us. The complexity and levels of calculation may vary but they do exist. This proves that even primitive humans were using their mental faculties for the purposes effective for them at that point in time. So how did it evolve to the present time? Well, it can be said that initially human beings were living a Stone Age life. Their main concern was to survive and fetch food just like the basic functions of the various animals and living beings around them. They were, however, curious and while it cannot be said with precision when exactly did they start conscious decision-making, but they did it for sure.

History—II

The initial achievements of mankind were the tools made for hunting and the discovery and use of fire for warmth and security. This showed their calculating mind. Whatever little communication was there was through sign language and grunts or gestures. The big leap in human evolution occurred when they started farming and settled at one place. This gave them some stability and leisure time to think about things other than basic survival. Slowly, newer tools were developed and languages came into being. As humans advanced to different ages such as the Iron Age and the Copper Age, their tools became more sophisticated. Now, at which point in time did calculation start?

It can be presumed that as humans started settling, farming and herding cattle across the world and in the Indus valley, Mesopotamia and Egypt near river beds where there was enough water, human culture and civilization started to slowly evolve. People started building things such as tools, houses, building weapons, etc. There must have been a time when things were peaceful. Then there emerged the original sin of jealousy and people started comparing with each other. Then there started a race to amass worldly possessions. People started robbing and stealing from each other. Then there emerged the need for protection. So some people became protectors and started defending others and assumed the powers of a king.

All this evolution was slow but certainly evident. While all this was happening did some people decide that they would develop mathematics? While initially it did not happen that way, slowly there emerged the need to keep account and calculate ones possession. That need was the first step in the road to all the marvellous discoveries by mankind over the ages.

People who had wealth wanted to keep track of them and counting them with the help of human symbols such as by counting on the fingers was not sufficient. As the need increased, so did the pace of mathematical evolution. This, therefore, was the birth of mathematics. There were discoveries and inventions not localised to one region. As the human population dispersed over the globe, evolved, mingled and reproduced and as newer civilizations came into contact with each other, it became obvious that mathematical reasoning was universal, logical and need based.

Not only was mathematics evolving, it was evolving all across. While we cannot say who invented it first, every ancient civilization contributed to its evolution indicating that mathematics was a part of human life wherever it thrived and prospered.

Mesopotamian civilization: Mesopotamia is the land between the Euphrates and the Tigris rivers, both of which originate in the mountains of Armenia in modern-day Turkey. It was called the land between the rivers and also the cradle of modern civilization. Many artefacts were discovered indicating the Bronze Age and life in general in those times. The usage of mathematics in the Mesopotamian (Also called Babylonian civilization. Since we later refer to it as Babylonian civilization)civilization is proved from the clay tablets unearthed from that area during excavation. The clay tablets covered topics such as fractions, algebra, quadratic and cubic equations and the Pythagorean theorem.

We have a decimal system with (base 10) number system. However, the Babylonian system of mathematics was sexagesimal (base 60). From this we derive the modern day usage of 60 seconds in a minute, 60 minutes in an hour, and 360 degrees in a circle. The Babylonians were able to make great advances in mathematics for two reasons.

First, the number 60 is a superior, having divisors of 1, 2, 3, 4, 5, 6, 10, 12, 15, 20, 30, 60 (including those that are themselves composite), facilitating calculations with fractions. Additionally, unlike the Egyptians and Romans, the Babylonians had a true place-value system, where digits written in the left column represented larger values (much as in our base 10 system; for example, 734 = 7 × 100 + 3 × 10 + 4 × 1). The Sumerians and Babylonians were pioneers in this respect."

Babylonians knew the basic rules for measuring volumes and areas. The volume of a cylinder was taken as the product of the base and the height. The Pythagorean theorem was also known to the Babylonians. Also, there was a recent discovery in which a tablet showed used π as 3 and 1/8. The Babylonians are also known for the Babylonian mile, which was a measure of distance equal to about seven miles (or 11.3 kilometers) today. This measurement for distances eventually was converted to a time-mile used for measuring the travel of the Sun, therefore, representing time.

Thus it shows that Babylonians or Mesopotamians were very much aware of their surroundings and using whatever resources was at hand to get an understanding of their life and surroundings.

Mayan Civilization: This is also called the Mesoamerican civilization. Located in South America and in present day Mexico, this civilization thrived much before the advent of Columbus and the Spanish acquisition. **The** Mayans developed a very sophisticated version of mathematics. The Mayan counting system has only three symbols: A dot representing a value of one, a bar representing five, and a shell representing zero. These three symbols were used in various permutation and combinations so that the most ordinary man could understand and use them in his daily life. The only other civilization other than the Indus Valley civilization to have understood, and used the concept of zero is the Mayan civilization. This is remarkable as most other civilizations did not know the full importance of zero and how it could be used.

The Mayans used the vigesimal system for their calculations—a number system with the base 20 rather than 10. This means that instead of the 1, 10, 100, 1,000 and 10,000 of our mathematical system, the Mayans used 1, 20, 400, 8,000 and 160,000.

They used graphical numbers or glyphs as below:

| 0 | 1 | 5 | 18 | 20 |

The n set of mathematical symbols allowed even uneducated people to add and subtract for the purposes of trade and commerce. To add two numbers together, for example, the symbols for each number would be set side by side, then collapsed together to make a new single number. Thus, two bars and a single dot representing 11 could be added to one bar to make three bars and one dot, or 16.

The Mayans considered some numbers more sacred than others. One of these special numbers was 20, as it represented the number of fingers and toes a human being could use to count. Another special number was five, as this represented the number of digits on a hand or foot. Thirteen was sacred as the number of original Mayan gods. Another sacred number was 52, representing the number of years in a "bundle", a unit similar in concept to our century. Another number, 400, had a sacred meaning as the number of Mayan gods of the night.

There are various articles on the internet which can give a lot of information on the Mayan civilization and their usage of mathematics.

Aztec civilizations:

The Aztecs were the inhabitants of the modern-day Mexico. They were also a part of the ever-growing and invading civilizations which took over existing ones. Just like the Mayan civilization, the Aztec civilization it used pictorial representation to show its possessions and properties. The Aztecs used hand, heart and arrow symbols to represent fractional distances when measuring land. Whether they were used for calculating is debatable but we can conclude that mathematics was evolving and people were becoming proficient in counting as their land holdings increased.

Many old scriptures and graphical images depict their symbols in plotting their agricultural lands with symbols of calculation. When the researchers realized that the arrow, hand, and heart drawings represented ground distances, they were finally able to arrive at the same calculations as the Aztecs. Each symbol stood for a distance that was less than the standard distance unit, called a land rod.

The Aztecs had their own form of arithmetic. Just like the Mayans they used a base-20 number system, also called vigesimal (base 20) number system. A single dot ($\bullet$) represented the number 1. Occasionally, a finger was also used to indicate the number. The numbers 2, 3 and 4 were represented by two, three, and four dots, respectively. The number 5, though, introduced a new symbol; it was represented by five dots ($\bullet\bullet\bullet\bullet\bullet$). From 6 through 9, dots alone or a combination of dots and a bar represented the respective numbers. The number 10 was represented by a rhombus, two bars, or ten dots.

African civilization:

Africa is a vast continent and is one of the last few vestiges untouched by modern civilizations. There are many tribes and cultures prevalent and several languages and dialects spoken. There ought to have been usages of mathematics as per customs prevalent in the region. A rich and thriving culture like that of Africa must have parallely evolved just like others and may have used mathematics, though rudimentary by today's standards, but quite unconsciously relevant to its times.

The Lebombo bone is the oldest known mathematical artifact found dating back to the Paleolithic era. Again, the Ishango bone is a bone tool, dating back to the Upper Paleolithic era. It was first thought to be a tally stick, as it has a series of tally marks carved in three columns running across the length of the tool, but some scientists have suggested that the groupings of notches indicate a mathematical understanding that goes beyond counting.

The Nile valley of Egypt is also located in Africa but I will talk about it later. All of the mathematical findings of the Islamic world during the medieval period were by Timbuktu (a place in African) scholars: Arithmetic, algebra, geometry, and trigonometry. One of the major achievements traced to Africa was the advance knowledge of fractal geometry and mathematics. The knowledge of fractal geometry can be found in different aspects of African life—art, social design structures, architecture, games, trade and divination systems.

ı	∩	৩	↑	↑	◠	⚚
1	10	100	1000	10000	100000	10^6
Egyptian numeral hieroglyphs						

Egyptian civilization: A very high form of mathematics was used by the people living in the ancient civilization of Egypt by the river Nile. The Egyptians were famous for their pyramids and mummies and their various hieroglyphics and pictorial graphs. Theirs is a decimal system just like the modern-day numeral system. There are representations on clay tablets to show their addition, subtraction, multiplication and divisions. The numerals looked like this

Egyptians used fractions too, and were well-versed in the knowledge of algebra and geometry. They knew the calculation of areas and volumes. These helped them in storing their food grains in the huge granaries and also built huge pyramids which we today recognize as one of the wonders of the world of all times. The colossal structures and the pattern of perfect geometrical alignment in building these huge structures in those ancient times without the use of modern technology is in itself proof of the use of a highly developed form of calculation and technology.

Greek civilization:

Greek mathematics refers to the mathematics written in the Greek language. Greek mathematicians lived in cities spread over the entire eastern Mediterranean region, from Italy to North Africa, but were united by culture and language. Greek mathematics of the period following the reign of Alexander the Great is sometimes called Hellenistic mathematics. Many modern-day mathematical theories have their origin in the mathematics that developed during this time. In fact, the word "mathematics" was coined during this time. There were many great mathematicians like of Pythagoras, Plato, Euclid and Archimedes who were among the great philosophers and logical thinkers of their times. They contributed immensely to the development of deductive reasoning and logical thinking.

Earlier thinkers used inductive reasoning, i.e., repeated observations were made to establish the rule of the thumb. These were then formed into axioms and principles on the basis of which logical conclusions and theorems were derived. So we can say that deductive reasoning was based on assumptions proved earlier based on inductive reasoning. These thinkers used several concepts such as Pythagorean theorem, integration, irrational numbers, methods of exhaustion to reach a variety of conclusions.

Euclid wrote the book "Elements" which included all the concepts of math of those time and which are also relevant today such as number theory, algebra, solid geometry etc. Attempts were also made to calculate the value of Pi and develop trigonometry during this period.

Thus we can say that the Greek era was a very flourishing period and starting point for many further discoveries to come, based on the discoveries and inventions made during this time.

Chinese Mathematics:

Early Chinese mathematics is so different from that of other parts of the world that it is reasonable to assume that it developed independently. China has been a continent in itself and can be proud of many things which it claims to be have developed independently as there were many old traditions preserved and which continue to survive to this day. Their mathematical counting is also called Rod numerals. The depiction given below is as sourced from the Wikipedia.

Counting Rod Numerals

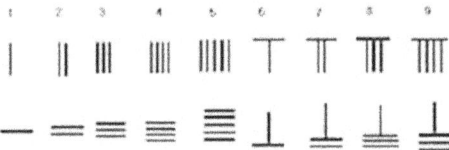

Of particular interest is the use in Chinese mathematics of a decimal positional notation system, the so-called "rod numerals" in which distinct ciphers were used for numbers between 1 and 10, and additional ciphers for powers of ten. Thus, the number 123 would be written using the symbol for "1", followed by the symbol for "100", then the symbol for "2" followed by the symbol for "10", followed by the symbol for "3". This was the most advanced number system in the world at the time, apparently in use several centuries before the Common Era. Rod numerals allowed the representation of numbers as large as desired and allowed calculations to be carried out on the *suan pan*, or Chinese abacus. They independently developed very large and negative numbers, decimals, a place value decimal system, a binary system, algebra, geometry, and trigonometry.

Even after European mathematics began to flourish during the Renaissance, European and Chinese mathematics were separate traditions, but Chinese mathematical output declined from the 13th century onwards.

Indian Civilization:

The earliest civilization on the Indian subcontinent is the Indus Valley civilization that flourished between 2600 BC and 1900 BC in the Indus river basin. Their cities were laid out with geometric regularity. This shows that the people of those times were proficient in mathematical subjects like geometry and knew basic calculations.

Many large numerals up to 10^12 were recited orally. They were not documented as they could have been destroyed, so there was a tradition to preserve the knowledge through oral recitation by the Brahmin community.

The concept of zero was first recorded and conceptualized by scholar Aryabhatt. Many scholars like Brahmagupta, Bhaskara-II, Aryabhatt, etc., made many contributions to mathematical concepts of zero as a number, negative numbers, arithmetic, algebra, trigonometry, sine and cosine. These mathematical concepts were transmitted to the Middle East, China, and Europe and led to further developments that now form the foundations of many areas of mathematics.

One of the more colourful figures in 20th-century mathematics was Srinivasa Aiyangar Ramanujan (1887-1920), an Indian autodidact who proved over 3000 theorems, including properties of highly composite numbers, the partition function and its asymptotics, and mock theta functions. He also made major investigations in the areas of gamma functions, modular forms, divergent series, hyper-geometric series and prime number theory.

However, there is a debate about it in the modern methodical world. The ancient temples and architectures are proof of a modern society with beautiful constructions and advance technological knowledge. The ancient Vedic scripture dates way back with descriptions collaborating Greek, Mesopotamian, and advanced mathematical knowledge, but there are no direct links to establish for want of proof. However, it showed a thinking society did exist.

As in China, there is a lack of continuity in Indian mathematics as significant advances are separated by long periods of inactivity.

History—III

Islamic mathematics: The Islamic period starts with the growth of Islam during the 8th century. Its influence was across Persia, the Middle East, Central Asia, North Africa, Iberia, and in parts of India Although most Islamic texts on mathematics were written in Arabic, most of them were not written by the Arabs, since much like the status of Greek in the Hellenistic world, Arabic was used as the written language of non-Arab scholars throughout the Islamic world at that time. Persians contributed to the world of mathematics alongside the Arabs.

Most of the concepts were already known but were further analysed and refined, and commentaries in the form of books were written. The scholars of those times were responsible for spreading the ideas to countries they came in contact with through trade and commerce.

In the 9th century, Persian mathematician Muhammad ibn Mūsā al-Khwārizmī wrote several important books on the Hindu-Arabic numerals and on methods for solving equations. His book *On the Calculation with Hindu Numerals*, written about 825, along with the work of Al-Kindi, were instrumental in spreading Indian mathematics and Indian numerals to the West. He studied an equation for its own sake and "in a generic manner, in so far as it does not simply emerge in the course of solving a problem, but is specifically called on to define an infinite class of problems".

Further developments in algebra were made by Al-Karaji in his treatise *al-Fakhri*, where he extends the methodology to incorporate integer powers and integer roots of unknown quantities. Something close to a proof by mathematical induction appears in a book written by Al-Karaji around 1000

AD, who used it to prove the binomial theorem. Mathematics historian, F. Woepcke praised Al-Karaji for being "the first who introduced the theory of algebraic calculus". Also in the 10th century, Abul Wafa translated the works of Diophantus into Arabic. Ibn al-Haytham was the first mathematician to derive the formula for the sum of the fourth powers, using a method that can be used for determining the general formula for the sum of any integral powers.

In the late 11th century, Omar Khayyam wrote *Discussions of the Difficulties in Euclid,* a book about what he perceived as flaws in Euclid's *Elements,* especially the parallel postulate. He was also the first to find the general geometric solution to cubic equations.

In the 13th century, Nasir al-Din Tusi (Nasireddin) made advances in spherical trigonometry. He also wrote on Euclid's parallel postulate. In the 15th century, Ghiyath al-Kashi computed the value of π to the 16th decimal place.

Other achievements of Muslim mathematicians during this period include the addition of the decimal point notation to the Arabic numerals, the discovery of all the modern trigonometric functions besides the sine, al-Kindi's introduction of cryptanalysis and frequency analysis, the development of analytic geometry by Ibn al-Haytham, the beginning of algebraic geometry by Omar Khayyam and the development of an algebraic notation by al-Qalasādī.

From 15th century onwards with the rise of the Ottoman Empire and the Safavid Empire, the development of Islamic mathematics stopped.

European mathematics:

In medieval Europe, religion ruled the masses. The religious leaders had succeeded in establishing the world order where very few people questioned things and it was left to the royal elites or Church who were concerned about searching for answers to various questions and at the same time about maintaining their authority. God was the answer to all difficult questions and ensure a chaos-free world. One driving element was the belief that mathematics provided the key to understanding the created order of nature, frequently justified by Plato's *Timaeus* and the biblical passage (in the *Book of Wisdom*) that God had *"ordered all things in measure, and number, and weight"*.

The subjects that were studied were arithmetic, geometry, astronomy, and music. They were relevant and sufficient for the time and supported the royal culture in the castles and courts. People were sometimes banished and declared heretics for going against the order of the times which was seen as the word of God. There were many instances where people such as Galileo and others were burned alive or hanged or guillotined for holding a different point of view of the world or universe and its surroundings.

In the 12th century, European scholars traveled to Spain and Sicily seeking scientific Arabic texts, including al-Khwārizmī's *The Compendious Book on Calculation by Completion and Balancing*, translated into Latin by Robert of Chester, and the complete text of Euclid's *Elements*, translated by Adelard of Bath, Herman of Carinthia, and Gerard of Cremona.

These new sources sparked a renewed interest in mathematics. Fibonacci, writing in the *Liber Abaci* in 1202 and updated in 1254, produced the first significant treatise on mathematics in Europe since the time of Eratosthenes, a gap of more than a thousand years. The work introduced Hindu-Arabic numerals to Europe, and discussed many other mathematical problems.

By the turn of the 14th century, concepts such as motion theory which included components of force, resistance, speed, distance and time were discussed.

Euclidean mathematics was explored to find solutions to the various scientific problems of those times and laid a base for the invention of engines and vehicles and ship-building technology, etc.

Renaissance period mathematics:

With the advent of the Renaissance, all old concepts that till then had remained unquestioned were beginning to be challenged. This was the period which saw the emergence of thinkers who made new discoveries in their process of learning. Many scientists, mathematician, painters, sculptors and voyagers made learning their full-time obsession. Royal courts started giving grants to independent researchers and discoverers. This led to various discoveries, both material and theoretical. During the Renaissance, the development in mathematics and accounting were intertwined. While there is no direct relationship between algebra and accounting, the teaching of the subjects and the books published were often intended for the children of merchants who were sent to reckoning schools (in Flanders and Germany) or abacus schools (known as *abbaco* in Italy), where they learned the skills useful for trade and commerce. There is probably no need to know algebra to perform book-keeping operations, but for complex bartering operations or the calculation of compound interest, a basic knowledge of arithmetic was mandatory and knowledge of algebra was very useful.

Concepts such as trigonometry and sine and cosine formulae led many voyagers to calculate the distances with the help of planetary directions. During the Renaissance, the desire of artists to represent the natural world realistically, together with the rediscovered philosophy of the Greeks, led them to study mathematics. They were also the engineers and architects of that time, and so needed to know mathematics in any case. The art of painting in perspective, and the developments in geometry that involved, were studied intensely.

The 17th century saw an unprecedented explosion of mathematical and scientific ideas across Europe. Invention of logarithms helped develop mathematical laws of planetary motion. The analytic geometry developed allowed those orbits to be plotted on a graph, in Cartesian coordinates & modern decimal notation capable of describing all numbers, whether rational or irrational.

Building on earlier work by many predecessors, English scientist Isaac Newton discovered the laws of physics and brought together the concepts now known as infinitesimal calculus. Independently, Gottfried Wilhelm Leibniz, who is arguably the most important mathematician of the 17th century, developed calculus and much of the calculus notation still in use today. Science and mathematics had become an international endeavour, which would soon spread across the entire world. In addition to the application of mathematics to the studies of the heavens, applied mathematics began to expand into new areas with the correspondence between Pierre de Fermat and Blaise Pascal. Pascal and Fermat laid the ground work for the investigations into probability theory and the corresponding rules of combinatorics in their discussions over a game of gambling.

The most influential mathematician of the 18th century was arguably Leonhard Euler. His contributions range from initiating the study of graph theory. He made numerous contributions to the study of topology, graph theory, calculus, combinatorics, and complex analysis, as evidenced by the multitude of theorems and notations named for him.

Other important European mathematicians of the 18th century included Joseph Louis Lagrange who did pioneering work in number theory, algebra, differential calculus, and the calculus of variations, and Laplace, who during the rule of Napoleon, did important work on celestial mechanics and on statistics.

Modern-day mathematics:

Modern-day mathematics can be said to be contributing the advanced mathematics based on the Basic Mathematics of earlier times and having its usage in applied mathematics a new classification to contribute all the discoveries to the application of known mathematics. Many people from across the world have contributed significantly to the development of mathematics. The use of computers and discovery of groundbreaking theories have led to exponential growth in the realm of mathematics. Concepts such as complex variables, in geometry, convergence of series, fundamental theorem of algebra, the quadratic reciprocity law, non-Euclidean geometry, abstract algebra, non-commutative algebra, Boolean algebra, parameter space and hyper-complex numbers have all led to further development of mathematics.

The 20th century saw mathematics become a major profession. Every year, thousands of new PhDs in mathematics are awarded, and jobs are available in both education and industry. An effort to catalogue the areas and applications of mathematics was undertaken in Klein's encyclopedia.

In the year 1900, in a speech to the International Congress of Mathematicians, David Hilbert set out a list of 23 unsolved problems in mathematics. These problems, spanning many areas of mathematics, formed the central focus for much of 20th-century mathematics. Today, 10 have been solved, 7 are partially solved, and 2 are still open. The remaining 4 are too loosely formulated to be stated as solved or not.

Mathematical collaborations of unprecedented size and scope have taken place. An example is the classification of finite simple groups, also called the "enormous theorem", whose proof between the years 1955 and 1983 required 500-odd journal articles by about 100 authors, and which filled up tens of thousands of pages.

Differential geometry came into its own when Albert Einstein used it in proving the theory of general relativity. Entire new areas of mathematics such as mathematical logic, topology, and John von Neumann's game theory changed the kinds of questions that could be answered by mathematical methods. All kinds of structures were abstracted using axioms and given names such as metric spaces, topological spaces, etc. The concept of an abstract structure was itself abstract and led to category theory. Lie theory with its Lie groups and Lie algebras became one of the major areas of study.

The development and continual improvement of computers, at first, mechanical analog machines and then digital electronic machines, allowed industry to deal with larger and larger amounts of data to facilitate mass production and distribution and communication, and new areas of mathematics were developed to deal with this.

Classification of Mathematics

Classes of mathematics: Mathematics was much simpler earlier, but it has grown very vast today due to the inclusion and involvement of various disciplines and sub-disciplines or areas of subjects which were either discovered by using mathematics or are using maths for the first time. Thus we have virgin fields yet to be explored and fields which have been touched upon as a way of romancing the known with the unknown in pursuit of curiosity and happiness. Despite the inter-disciplinary development of mathematics and its exponential growth, the subject can be classified into two broad groups: Basic Mathematics and Advanced Mathematics.

The basis of classification is as per the historical evolution of mathematics. Advanced Mathematics can be further sub-divided into: A) Pure Mathematics and & B) Applied Mathematics.

This classification is done to give a more current perspective of ancient concepts by way of refinements in Pure Mathematics and the application part (Applied Mathematics) shows all those classes that evolved to solve problems and use of various combinations for the purpose of using the knowledge of mathematics, both ancient and, for solving today's and future problems.

The classifications given below listing the various branches and divisions of mathematics has been done to make the reader become comfortable with the various terms and subheads used in mathematics. This classification itself can be debated, but it gives a much more rational picture of the humungous world of mathematics.

1. BASIC MATHEMATICS

1.1 Discrete Mathematics.

Discrete mathematics is the study of mathematical structures that are fundamentally discrete rather than continuous. Some concepts falling under discrete mathematics are as below

1.1.1 Sets
    1.1.1.1 Set(Mathematics)
        1.1.1.1.1 Element
        1.1.1.1.2 Venn Diagram
        1.1.1.1.3 Empty Set
        1.1.1.1.4 Subset
        1.1.1.1.5 Union
        1.1.1.1.6 Intersection
        1.1.1.1.7 Complementary
        1.1.1.1.8 Symmetric Difference
    1.1.1.2 Ordered Pair
    1.1.1.3 Cartesian Product
    1.1.1.4 Power Set
    1.1.1.5 Simple theorem in algebra of sets
    1.1.1.6 Naïve set theory
    1.1.1.7 Multi set

1.1.2 Functions
    1.1.2.1 Function
    1.1.2.2 Relation Composition
    1.1.2.3 Permutations
    1.1.2.4 Symmetry

1.2 Calculus
    1.2.1   Pre-Calculus
    1.2.2   Limits
    1.2.3   Differential Calculus
        1.2.3.1 Derivative
        1.2.3.2 Notation
    1.2.4   Integral Calculus

1.3   Geometry
    1.3.1   Euclidian Geometry
        1.3.1.1 Foundations
            1.3.1.1.1 Hilbert's Axioms
            1.3.1.1.2 Point
            1.3.1.1.3 Locus
            1.3.1.1.4 Line
            1.3.1.1.5 Parallel
            1.3.1.1.6 Angle
            1.3.1.1.7 Congruence
            1.3.1.1.8 Similarity

        1.3.1.2 Plane & 2 Dimensional Geometry
            1.3.1.2.1   2D Geometric Model
            1.3.1.2.2   Altitude
            1.3.1.2.3   Cone
            1.3.1.2.4   Circle
            1.3.1.2.5   Parabola
            1.3.1.2.6   Hyperbola
            1.3.1.2.7   Ellipse
            1.3.1.2.8   Triangles
            1.3.1.2.9   Quadrilateral
            1.3.1.2.10 Polygon
            1.3.1.2.11 Trapezium

1.3.1.2.12   Trigonometry

1.3.1.3   3 Dimensional Geometry

1.3.1.3.1   3-D Projection

1.3.1.3.2   Hyperboloid

1.3.1.3.3   Paraboloid

1.3.1.3.4   Polyhedron

1.3.1.3.5   Spheroid

1.3.1.4   N Dimension Geometry
1.3.1.4.1   Ball
1.3.1.4.2   Convex

1.3.1.4.3   Parallelogram Law

1.3.1.4.4   Sphere

1.3.1.4.5   Root System

1.3.2   Non Euclidian Geometry
1.3.2.1   Arc (Projective Geometry)
1.3.2.2   Line at Infinity

1.3.2.3   Point at Infinity

1.3.2.4   Plane at Infinity

1.3.2.5   Non Euclidean Geometry

1.3.3   Others Forms out of Many
1.3.3.1   Algebraic Geometry
1.3.3.2   Complex Geometry

1.3.3.3   Differential Geometry

1.4 Logic
1.4.1   Foundational
1.4.1.1   Analytic

1.4.1.2   Definition

1.4.1.3   Inference

1.4.1.4   Name

1.4.1.5   Logical Truth

2. ADVANCED MATHEMATICS

Thus the list given above is just vast but not exhaustive and does not include all divisions and sub-topics. I have included topics which most of us might have heard about, but will certainly give us a clear idea as to how and where mathematics has evolved and is still evolving. These topics and some I have never heard of and don't know where they are used but someone has and some historian has made sure that it's recorded and classified. So if you feel side-lined and think you don't know it, don't worry, so do many of us. And if someone wants to take a dip in this vast pool and make a name on some topics don't worry again there is a lot going on and a lot can be further done.

Contemporary Issues

We now proceed to contemporary issues in the study of mathematics. We will deal with questions that have troubled more or less everyone in their growing years. I will try to answer each question on the basis of what I have learnt over the years and my analysis of the subject.

We will deal with each question separately and get answers to them as I interpreted them. These questions were introduced in the opening paragraph of my book.

The Big Questions:

What prompts us to learn mathematics in school, colleges and universities?

Many of us have never given it a thought as to what prompts us to learn mathematics. The obvious answer is that most of us never had any choice. In the modern schooling system, mathematics is a compulsory subject. There is never a choice as to whether we want to learn it or not. Right from the beginning, we are introduced to the subject with the view that it is used in daily life and we must know it.

This is the point from where we all start. However, after learning it as a mandatory subject in basic schooling, we do get a choice to opt out when we are in high school, college or university. This is when we make a decision whether to take up mathematics as a subject or not. In high school and college, learning mathematics has a lot to do with our career progression. For example, if you want to be a computer programmer it is important to know maths. Similarly, if you want to major in engineering or technology, you ought to know relevant topics and concepts. So it's more to do with personal choice as to what one wants to major in and then to do that whether one needs to know maths. But then the question is why do we decide to choose a subject that needs a maths foundation and advanced knowledge of that subject?

This can be either due to genuine interest or pressure from peers and parents. I may have a goal to be a scientist or a mathematician or an engineer or a technician. Every parent expects his or her child to excel in education and it's only by doing that that one is able to choose a desired profession and then land a dream job. Parents force their children to get tuitions and professional help right from middle school so that their base in mathematics is sound. This need or anxiety driven greed has a lot to do with the aspirations and needs of parents and peers alike. Parents want to have children with excellent mathematical skills as it is associated with being smart, intelligent and wise. This is ultimately associated with success.

Every child wants to be like his or her friends or peers and not look like an oddball. So if all my friends are taking mathematics as their major subject and want to join a computer programming course, I also would be inclined to follow them.

Here, it's more to do with the need to be accepted and be associated with a group.

Sometime this creates a race to be the best and learn this subject. One is prompted to know it so that he or she has the option to go to the best universities.

This may not be the only reason why one is prompted to learn mathematics learning but can be a contributing factor.

Why are parents and educators losing their sleep over their children's/ students' performance in mathematics?

While the answer to this has been touched upon in the previous discussion, it has a lot to do with the aspirations of the parents and teachers. Parents want their children to excel in life. They have seen that the median salary of the person who is an engineer, doctor; technician, scientist or a mathematician is well above the average. Those who are good in mathematics are assumed to do well in life and are good in calculations as well, while the two are actually quite different. Also, those parents who have their own business want their children to be good in mathematics so that they can take care of their business better. The number of jobs, scholarships and grants given to students who excel in mathematics is much higher than those given for other fields. This gives a vision of success and a fulfilling life ahead.

So the parents want their children to learn maths right from childhood and excel in it so that they can always be in the lead and have a successful life ahead.

When it comes to teachers, their matrix of performance is driven by the number of students successfully clearing their class. The greater the number of students passing and getting better grades, the better is the personal prestige and professional growth of the teacher. Sometimes parents pressurise the teachers to give extra attention which creates additional pressure on the teachers as well. The educational boards that set the syllabus also compete to ensure that the number of graduates passing from the reputed schools, colleges and universities are on the rise. This has to do with the race to produce the best research papers, articles, inventions, discoveries, patents, etc. The country's government wants to compete with the rest of the world and wants it to be seen as a leader in cutting-edge science and technology. This has to do with national pride and world dominance and trade. Therefore, billions of dollars is pumped into the education system so that the country produces the best minds and best technicians, professionals, managers and scientists. Excellence in mathematics is required in each and every field.

This creates pressure on the teachers to ensure that a subject which is rational and abstract be made more conceptual by keeping its objectives alive and drilling the results through it. The difficulty is in the enormity of responsibility and the vast span of time that elapses from the time the subject is introduced to the time it will be applied. There are several unknown variables or factors which can come in the way till results are achieved at both the micro-level and at the macro-level.

This dependence and institutionalization of the subject has led to sleepless nights for both parents and teachers. Parents and teachers along with the students are a part of the society which has already accepted this subject as something magical and capable of producing exceptional results.

What happens when one doesn't know maths?

Well, the answer is a little different for every country, society and community. However, it does create an impact. For example, if a country is not solely dependent on the number of scientists, engineers, computer programmers and technicians it is churning out as an indicator of its progress and as a matrix to display its superiority and leadership in the world, it will have a subtle if not direct effect. Out of all the countries in the world keeping everything equal the developed countries are able to maintain their leadership due to the huge number of new technological developments, discoveries and inventions they have made. It is through patents and proprietorship of these that they are able to bargain with the rest of the world. If that comparative advantage or bargaining power is lost, it can directly affect its trade, commerce and competitive advantage over others.

Therefore to say not knowing maths will not make a big difference to the people of those countries that don't trade their technical skills with the rest of the world is a fallacy.

Well . . . . almost as they can always produce writers like me (laughs), painters, sculptors, artists, managers, and people who rely on other sciences and arts to bargain with the rest of the world.

However, even in contemporary fields where mathematics is not essential, the mathematical and digital revolution has touched and influenced their growth. This has made this subject very important from an idealist or leadership point of view. Countries and people have survived and even flourished without consciously been associated with the subject and will do so in the future too. However, their relevance will diminish as more and more people rely on the subject.

Here I have looked at the effect on world trade, economics, business and leadership.

But what about the effect it has in one's life. Well, again this is quite relative as to how one leads one's life.

There are times when friends, parents, teachers, colleagues and acquaintances may reacted uncaringly not knowing the individual sensitivity of each student. Some may take a negative feedback without much thought and continue with their life without any self-doubts, while others may start believing that they are genuinely weak and dumb in mathematics and would not be able to make any progress in a career related to that subject. This can be quite shattering for a person who may have dreamt at one time to pursue a vocation that was connected to mathematics in some way. This could well be the reason why one chose not to study mathematics and gave up much before he or she really explored whether it was beyond their capacity.

In countries where there is a wide gap in opportunities for growth, jobs and education, not knowing maths can be very costly and can have profound long-term effects on one's choice of career and thereby prosperity. This is because all the good jobs in such countries depend on mathematics, and not knowing or rather not excelling in it can cost one a decent education from a good institution and the opportunity to get a good job and thus improve his/her status in life.

These consequences are only in those countries where there is too much reliance on technical subjects for getting good jobs.

Even if we keep all factors constant, lack of knowledge in mathematics can probably make one a little more vulnerable in terms of making decisions which need basic calculation and make one susceptible to fraud. However, if third party service providers such as sincere consultants and advisors are available this risk can be reduced considerably.

What is the teen and peer pressure associated with excelling in mathematics?

Learning mathematics is like learning other sciences but with a certain abstraction associated with it as there are certain things which one can safely assume. Each of us has a different personality as per our family background, culture and beliefs. Some learn to succeed and get their job done. They accept things as they are and carry on with that understanding without getting stuck in the details, whereas there are others who want to get into the depth of things and are not satisfied with the answers given to them. So while your friend may be able to understand a concept quickly due to his special circumstances and to an extent his logical and analytical bent of mind, you may not be on the same track and fall behind.

I would like to add that this definitely does not mean that you are less intelligent or smart than your friend but one is under tremendous pressure as a child to keep up with his peers. The teacher has to cover a lot of topics in a limited time frame, so she too has to rush forward irrespective of whether everyone understands each concept or not. Over the years the stock of knowledge has only increased and the syllabus is constantly churned to keep abreast of the latest findings. Does this mean that the students have become smarter as their heads are stuffed with more facts in the same amount of time or has the years of schooling doubled from what they used to be as the field of learning has multiplied?

Neither seems to have happened, so what this means is that either some students assume they have understood and move ahead without worrying about the details and smartly score in their exams, while some question everything and due to that disposition get lost. None of the approaches are wrong from the learning point of view if one knows what is happening. It's here that a good teacher can pull up someone who has the necessary aptitude by helping them change their perspective a little just like a sports coach does, while get someone else to focus on the details which may have been overlooked.

Even after these broad personality classifications, there are some people who are genuinely a little better than others due to their sheer energy and some who simply have a different and sometimes challenging disposition. In case of the latter, it is better to be satisfied with the best from the kid rather than pressurise him to perform like his peers. Remember, learning mathematics and being good at it is not proof enough that one is better than others and is going to be successful for sure. It can only mean that the person stands a better chance in a situation which is pro-mathematics.

Mathematics is just a part of the vast universe of philosophies and subjects found. One can certainly excel in many other subjects in which he or she is interested.

How is mathematics affecting employers and job seekers?

The study of mathematics is growing at a very fast rate. Every field that it comes in contact with probably can benefit from it. This has lead to several new fields being opened up for further study and research. This can be seen by the various doctorates and PhDs given by several universities across the world for path-breaking new discoveries made in the field of mathematics. In fact, the application of mathematics in other disciplines is studied under a separate field called Applied Mathematics. This shows that universities approve of students doing research in mathematics and grants are made by multinational corporations and trusts for further work in the field. This association of the corporate world with schools, colleges and universities shows a healthy trend for further job creation and the growth of the subject as a whole.

I will explain how this whole process works. Say, for example, a big multinational company in the electronics business is funding a university for its doctoral subjects. These universities will select certain students who have outstanding achievements to their credit and are interested in working on the subject. Now doing a PhD is a self-learning and exploratory process. Also, no two doctoral subjects are same. This ensures diversity of subjects with doctoral credit and work is done on multiple fronts. Now if the student discovers a better technology in the same field in which the company functions it will pave the way for patents and new technological rights. These rights will be sold to the same company who will then employ the student to pioneer the launch of the new product with the latest technology developed.

This is a win-win situation where the economy and the industry are benefitted by better technology created, the consumers get better products and innovation is encouraged. The job-seekers get jobs as there is a spurt in demand for engineers and technicians having the latest technological knowledge. The companies get better returns on their investments in the trusts and colleges or through their corporate social responsibility work. They recover the cost incurred by developing a fresh market, repeat orders and shorter product lifecycles.

Thus employers and job seekers benefit from the role of mathematics being a catalyst in revenue generation, capital formation and intellectual property being generated. This enables companies to create product differentiation and enable a steady supply of innovative products and satisfy the needs of the consumers. The consumers get better value for money. The government gets a better trade balance and its trade deficit is reduced. If the company is a multinational, then by improving exports it earns valuable foreign currency and reduces the current account deficit. The money spent by the government to improve the infrastructure by enabling university grants allows it to maintain its leadership position in the world in the given product or technology.

Thus all the stakeholders benefit from mathematics.

What is the effect of mathematics on business?

The answer to the above question has already been discussed in the previous question but I would like to add that mathematics touches everyone who comes in contact with it.

Business is not an independent unit of an economy. It is a system of producing, maintaining and further generating capital for the society at large. We have seen in the previous discussion how a corporate house invests in a university doctoral program and in return gets state-of-the-art technology for its product the rights to which it initially buys and then uses it to manufacture products on a mass scale and get a return on its investment.

Those businesses which do not benefit directly from a change of technology will have to improve their competitiveness by adopting the same technology or will have to develop newer technologies or substitutes which are in demand. They will have to create a niche through the same means or by improving their service, supply chain, backward and forward linkages in the production chain, and create a differentiation. This will lead to improvement in other areas which were neglected. They have to change or they will become extinct. Thus a change in one area in a particular industry will create a positive wave of changes and lead to efficiency in the market. This will propel the government to create infrastructure to meet the demands of industry and thus improve competition. The end result will be better and efficient markets are created.

It's not a zero sum game but one where the sum of the parts is greater than the parts themselves.

Thus the effect on business is compounding even if it does not directly participate and is not involved in any technological breakthrough.

What's in it for anyone who comes in touch with mathematics?

All stakeholders as well as non-stakeholders benefit from mathematics. This has been discussed in the previous questions. As mathematics is universal in its presence, it appeals to everyone who comes in contact with it. As we have seen through our discourse, that right from the primal times till today, mathematics had and has influenced us directly or indirectly through its presence and application.

When we are using a device which uses digital technology, we are being touched by mathematics. Our banking system, trade and commerce, technology, electrical systems, education system, political system, the computer which we use, the programme written for it, the hardware all of these have been influenced by mathematics.

Now whether we know the algorithms behind a computer programme or a digital technology, we have definitely benefitted from its applications. So even if you do not know the power of computing, you will see its benefits in your life and consider it as a part of your life. That is why we need to know how and where mathematics is touching us in our daily lives. Once we know its omnipresent existence, we will stop fearing it as a subject and start appreciating it, try to know how we can use it and even explore it for further learning, research and understanding.

What is the solution and how should mathematics be taught so that children look forward to it?

The solutions are many. Mathematics can be taught by including various different methods or it can be taught by simply continuing with the old schooling methods. This need for how to teach this grand subject itself is the problem which we have to solve. Mathematics is not just the teaching of a few textbooks or chapters as it used to be when it was first studied. The subject, just like everything else, has grown manifold. So cramming up the facts within a limited time period is not the solution. Educationists often include concepts which will form a base for modern-day applied mathematics.

Also, once a learner has understood his ability and interest in mathematics after studying basic mathematics for a few initial years, he can decide as to what course or stream he wants to pursue. Remember time and mind is limited, the spirit to learn is not. Also, it would take years before someone can decide what's best for him or her. However, the role of an educationist should be that of a facilitator who will guide, facilitate and create the right environment for the students to get basic knowledge in a fearless, uninhibited and dedicated atmosphere. If he understands the student and his background he will probably be able to direct his student much better. However I feel the people who are or will be made responsible for dissipating this knowledge should not balk at unlearning accepted ideas and should always be willing to go ahead with newer ways of teaching and learning.

Most of the time, the difficulty is that the task at hand seems to be huge and gigantic. The need to force or prepare a future takes some anxiety and conscious determination from the administration, but one should never forget the end user and ensure that this assimilation is done in a manner that not only develops a positive atmosphere but an atmosphere where negative suggestions are minimised, if not eliminated fully, and the students choose a field where trying and failing is accepted to be as normal as being successful.

How should the process of learning be initiated?

Learning is a life-long process. Also, it is not a destination but a continuous flow and a journey towards the goal of actualization. Learning can be initiated at any time and any place and by any one. One does not need any invitation to initiate the process. It also depends upon the situation of the learner. If the situation is adverse and the learner wants to learn, then one has to move towards the learner. However, if the situation is conducive and there are choices and options available, the learner should move towards the goal provided he is interested in acquiring more knowledge.

Many a time in adverse situations, it is difficult to collate the resources and get the necessary tools required to teach the subject effectively. Also, even if the situation allows learning, whether it is the quality of education or just plain understanding which lacks is debatable. Many a time things are often taken at its face value but the inherent understanding is lacking. In such a situation, administrators and educators find it challenging to facilitate and initiate actual learning.

This initiation of good learning practices can be done by active participation on the part of the learner and the one who imparts learning. There has to be mutual interaction so that it leads to healthy curiosity and acceptance. That means that even if the person makes mistakes and fails, his failure should be accepted without negative feedback and it should be made clear that whatever the person thinks or resolves is a stepping stone to his or her journey towards acquiring in-depth knowledge about the subject. Initial success or failure should not be the basis for determining whether the person has the necessary talent for the subject.

A good teacher will only reinforce the strong points and try to minimise the negatives and this will lead to a good initiation of the process of true learning. This will allow a healthy atmosphere where the true learning spirit is unleashed. When the mind is free from fear, be it of rejection or criticism, will it flow in the right direction and eventually seek its own level.

What is the future associated with Mathematics?

Mathematics is an ancient subject with current relevance and future connotation. Nobody knows what the future has in store for us. But that does not mean that we cannot control its outcome to a limited extent. If we have properly learnt to use mathematics in the early child education years, it will bear results in terms of an informed and analytical mind later on in life. So, if a person is faced with a situation where he has to take decisions, he can do so with the help of the skills gathered and developed due to learning the subject in his early years.

The future of the subject is directly proportional to its current usage by various stakeholders, administrators and educators. If they use it judiciously with their eyes trained on long-term application of the subject, it will be a service to the subject in term of growth and further application.

We, as stakeholders, have to know our place in the scheme of things and get what belongs to us and what gives us value. The subject can be shaped as per the demand of a particular direction or a particular application that may be shaped as per the future challenges which may emerge. This can be done by a planning committee who plans for the future taking into consideration the way mathematics is used now and what may be its best applications as the problems may extrapolate in the future.

Even when I am writing this, there are reseachers, scientists, educators, philosophers, writers, historians, planners, administrators, governments, corporates, trusts, foundations, schools, colleges, universities and many more people at work are creating and developing this fascinating subject.

Why do some students excel in mathematics while failing to master the subject?

The success or failure in a subject is never how much you score or what grade you get. It is done only to rationalise competition and instil that spirit so that when you have limited openings and multiple applicants, you can short list them. In this country where population growth is very high, resources are scarce. This has led to a system or structure where everything is calibrated and compared with to get the best and also ensure justice is done to all. Our society itself has been benchmarked so that when someone gets ten on ten it is better than someone getting nine or a five or even less. That benchmark has created a rat race where instant gratification and success is valued more than the ability to solve problems. Well, this method helps initially but is detrimental in the long run for those who believe in the system and go by the rule book.

We find that many children are good in their studies initially but somewhere down the line they get left behind. They are called dumb, loser, failure as they don't score as high as those who succeed and get the relevant grade. While that does not mean that we take away what the person has genuinely earned but it also means that everyone is different and the person who is better at some other subject must be equally respected and not criticised. Some students are genuinely smart and have a better understanding, but it's really difficult to say whether that means that they are better in all respects. Everyone has a purpose in life and this understanding itself will go a long way in alleviating the confused student who has an inferiority complex.

Mathematics is a rational and abstract inference and deduction based subject. There may be some students who have understood what to pay attention to and have been taught how to do it. Others who are not that so lucky will have difficulties. Whether initial success or failure is important is a different debate but we do find a complete disconnect between those who reject the subject outright and those who choose to have a try at it irrespective of its outcome. Failure makes us uncomfortable and humans want to avoid situations where there is pain or discomfort. So we find self-imposed suggestions, coupled with popular stereotyping responsible for creating an image for the person so that he avoids learning it and goes away from it further. Similarly, those who have tasted initial success may be led to believe they are special when they may be just as hard-working and ordinary as others and may have a superiority complex which may cause difficulties in future. So being assertive without being critical and encouraging eustress (positive stress) is a better way of learning rather than discarding mathematics altogether.

Question & Answer Series—I

In this chapter we will discuss the various questions being raised by educators, teachers and parents as to what we as guardians, parents, students or teachers feel about a particular situation. These questions indicate the current situation and what is happening at the grassroots. While attempting to answer these questions, I not only learned about the problems but it also made me understand how to approach these situations and how we can accept mathematics in our lives.

The questions may or may not be relevant to your situation but they will definitely bring a new perspective and provide you with an opportunity to ask similar questions for related situations. The discussion includes many suggestions and solutions which may help a parent or an educator if they face a similar dilemma. It may even give you answers which you never thought were possible.

Question No. 1:

Why do you find some people better at mathematics?

Response

I am not at all surprised because I totally agree given the way maths has been projected to everyone around. There seems to be always some people surprisingly better at deciphering facts effortlessly while some keep struggling. I myself was good at maths up to a point but somewhere after grade 4, I got lost as the facts multiplied and the links which I used to understand the subject became blurred. The whys and questions just increased. I think maths has evolved a lot with time and every fraternity has evolved their concepts differently, but it is more to do with applying means towards learning and teaching. Also, if one is curious and has questions then it becomes difficult to accept some laws as they are. The mind is trained not to ask too many questions, accept facts and proceed based on the assumption. The problem arises when the questioning mind sometimes gets in the way and it becomes difficult to accept things at face value. Also, the concepts taught at the basic level are sometimes the most advanced and require years of rigorous learning, reading and experience to have the whole picture unfold to its final applicability.

The best way to understand maths is to understand its implicit workings. If that's not possible, just accept the facts as in the rules of a game unquestioningly and then, as you grow ask questions at the right time, and you will probably get the answers.

Question No. 2:

What can be done to counter situations where students get stereotyped?

Response

Stereotyping is casting a particular person in to a particular class or group as per its similar beliefs or characteristics. Many a time we see people act as per heresay and common gossips without giving a serious challenge to the accepted thought. For Ex. If maths is considered to be not one's cup of tea and geeky, obviously people who fail initially will either give-up or considered to be geeks themselves. These acceptance of behavious as per pressures may not be relevant to the true capacities of the individuals. The person may be led in to believing things which may be far from truth and remain unchallenged.

In order to limit the effect of damaging stereotypes, the guardian or teacher or parent must make the student understand that at some point in life they will have to make decisions and they will have to trust themselves that they are making the right decisions. They will have to learn some really basic theories. This will propel better decision-making. The student's natural inclination may be to reject this advice, but by providing the best environment, lots of positive talk and countering negative feelings with an equally effective belief system will not only help instill self-confidence, but also make one a devout student of the subject.

Just as a particular child is labelled as dull and not so bright, there are children who are led into believing that they are smarter, and better at mathematics. The point here is that everyone is different. So how do you encourage the one who does good work and at the same time make sure they he or she does not get start assuming that he or she is better than others? Similarly, those who are struggling to understand the concepts should not be exposed to negative comments like "maths is difficult" or "You are not good at mathematics".

This conscious or unconscious stereotyping by teachers, peers and colleagues leads to a creation of groups and belief systems which may be natural but may lead to poor self-esteem. So what should be done? There is nothing wrong in celebrating success and smartness to motivate them to do better, but equal attention should be given to non-performers as well to bring them at par. If they face difficulties then they should be made to understand that it's alright to take some time to succeed.

The process of making mistakes itself should be encouraged so that the mind starts questioning without any fear of being stereotyped.

There have been experiments where people of similar intelligence are grouped together to get the best benefits, but I believe the mixed group is the best option, as in the real world one may have to deal with different kinds of people. So the earlier one gets an opportunity to interact with all kinds of people, the better it is for their confidence and personality development.

Question No. 3:

Why or how do you think negative or positive messages can have a long-term impact on a student?

Response

It's been often seen that in the formative years, young minds are pliable to acceptance without much retrospection and experience to fall back upon. This leads to children believing what is said about them unless one has a very solid and firm constitution and support group advising him or her. An inexperienced person may also achieve great things if backed by powerful self-belief. This can be true as persons who do not analyse their actions too much may be much clearer in their thoughts and use their imagination to achieve greater success. Many a time even a lie is believed to be true and is remembered for a long time after it is said.

Some people unlearn or relearn from friends and groups and reaffirm their messages, while others withdraw and stick to the world of books. This may bring different results depending upon the future environment of the subject and his own acceptance of the world view. If there are positive influences, he/she will remember that one message even years later. If there are negative influences and the child accepts it and changes his life pattern, obviously the message may be forgotten or overwritten by newer experiences, but it may still trouble him at some other time.

Question No. 4:
Suggest Activities to Help Learn Math

Response

Learn in small bits and pieces and then link it to bigger concepts. For example, suppose the child is learning about ratios. Give the introduction, go to its history, its present use and future application but keep the elaborate explanation for later. Give a brief introduction to catch the child's imagination. Interact with the students so that they are able to understand better. Give feedback and clarifications.

Explain the broad types, give genus or systematic classification and expand upon each class separately. Teach a concept by giving examples, allow them to practice, encourage sharing of notes, reward successful solutions, provide feedback, and repeat the process if required.

If the children still find it difficult, reassign the task, regroup and pull the ones still struggling back into the flow. If one student is lagging behind the others, have a one-on-one session if it is permitted. If it still does not work, it is okay to be an optimist and always be ready to help. The idea is to build confidence as much as possible.

Question No. 5:

What is that one thing you want to change in a classroom?

Response

There should be freedom of expression while experimenting with concepts and their application. But the problem with maths is we all follow a standard. That it is not changing according to our liking may seem like a restriction. The advancement in mathematics based on universally acceptable concepts creates a rat race and a system of gradation and rankings. Our achievement, success, and finally our position and happiness in life depend on it. This, therefore, makes us slaves to the system that we must follow. Many dropouts have performed well in computer science where they followed binary number system to programme an altogether new language. They were successful as they were allowed to learn creatively, and follow their natural curiosity at their own pace. This openness and creativity is sometimes missing in the classroom and needs to be changed.

There seems to be pressure to follow what others have discovered. But in the process, one loses that originality and curiosity that is possible only when one learns by discovering things oneself. Moreover, a follower has a lower sense of achievement and his satisfaction is less if he is not an excellent follower and follows things blindly. So one thing that needs to change in the classroom is to allow diversity of ideas while ensuring that they are in line with accepted standards, and explaining why it is done so that the student understands the importance.

And if they don't follow, well, let them re-write the laws if they can. After all, who can stop a new discovery?

Question & Answer Series—II

Question No. 1:

Where do you think that kids get negative messages about mathematics?

Response

Well, first and foremost, it is the joint responsibility of teachers, administrators, guardians, parents, peers and colleagues to not scare the students. When one is told that if he does not do well in mathematics then he has no chance in the competitive world, the child starts feeling the pressure to perform. Then there is re-enforcement from peers, friends, parents and others in general. Well it is true considerably as, there is a lot of dependence on maths. All the progressive contemporary sciences such as engineering, computers, advance learning depend on basic mathematics principles and if one is not good at them then one may not be able to get into these lucrative professions.

Most of the negative messages one gets are like: "Mathematics is a difficult and boring subject", "Mathematics is not my cup of tea", "I am not good at it", "If you are good at it you are a wizard", "If you don't know maths you can't expect to get into a good college, "You may have to drop out if you aren't clear about the fundamentals necessary for advanced technological subjects".

These and many other comments create anxiety, pressure, and belief that one cannot learn mathematics. This can rob a person of his/her self-esteem and his shot at a career which he/she might have opted for.

Question No. 2:

If schools decide to take the mindset evidence seriously, what would they need to change?

Response

If schools decide to take the mindset evidence seriously, then they will have to be careful in the choice of words while appreciating or praising the child. What they do casually may be interpreted differently and this may create differing experiences. However, I feel that the child may eventually learn from his experiences and does not have to be doomed for life due to one remark. It all depends upon what one wants to do in life.

However, when it comes to giving the best learning environment, it is best to inculcate the right values in a child and praise him for the effort he has made and try not to paint smartness as a God-given gift and instead appreciate it as a quality which is present in all. Some students are dreamers while others are active participants, so it will not be realistic to expect similar results from both. Nevertheless, it is important to appreciate good work and encourage non-performers to work better. It also depends upon what was the objective that the school wanted to achieve from that exercise, if the students have succeeded in achieving that, then that's a good job done. If not, then the school has to change its approach. The success parameters are then relative to the objectives needed to be achieved.

Question No. 3:

Why do you think mindset interventions impact girls and students of colour the most?

Response

The mindset interventions or stereotyping is the projection of one class or category of people with a certain attribute. Though the group in general may exhibit the attributes, it cannot be true of all and everyone in the group. These attributes may have been put through by media or interest groups who want to achieve a certain objective from the given class of people. Obviously, girls or women as a gender form a category in themselves and so do coloured students as their colour shows their ethnicity and region-centric classification.

These classifications are not needed as ultimately all are human. But we have seen that there are markets and politics associated with them and typifying them assures achievement of certain objectives. This creates a spiral of thoughts and typefication affecting girls and coloured students, but it's their individual reaction and decision which decides their future though it may be difficult due to the hype created.

Question No. 4:

Describe how the discussion has changed your thinking about your own ability in any way?

Response

The information in the discussion made me affirm my belief that mental growth is always built in the long run. That there are newer and stronger tracks created in the brain if one keeps on practising and continues learning, believing and reinforcing what he or she is learning. The brain is very adaptive and one's disposition, like positive mindedness and self-confidence, goes a long way in making one an affirmative and balanced individual. The person may not learn things overnight but it is okay to fail and accept losses in life and not be afraid of asking and be more open to suggestions and feedback.

If at all I have to teach a student I will probably be careful and think before I say anything which may have long-term consequences. One may offer words of encouragement provided the student is approachable.

The discussion and the information presented in the above pages can help one choose a path of measured approach and acceptance, so that the mind is calm and free from the mental baggage built up over years of neglect and negative comments.

Question No. 5:

How to think differently about students? And how can that change one's interactions with their children or students?

Response

We all have different levels of intellect and interests. If the objective of the parent is to ensure a particular skill-set in the student then he should not give up hope despite initial hurdles. This may help him to make an informed decision in the future. I have understood after years of analysis and an inherent curiosity as to why some people succeed academically while some don't when they seem to be doing fine in everything else. It's sometimes a matter of igniting the imagination and instilling belief in oneself which can propel a student to achieve excellence.

It is better to learn from one's mistakes then to surrender to failure. This kind of behaviour will only make a student think that a certain aspect is not worth understanding and he or she ought to avoid it as it make him or her feel vulnerable and weak. It is better to accept one's inability to understand certain facts, because only when you accept failure will you outgrow it. This will help you get a closure, stop the panic attacks and prepare yourself for future learning and success.

Question No. 6:

Based on ones experience as a parent or teacher, how does a student approach a mathematical problem? (Reader's are encouraged to think from their own experience)

Response

When there is problem the student stops and struggles with it. They want a step-by-step reason as to why there was a certain formula or rule followed and how the answer evolved. They expect the teachers to give them the answers. They may feel compelled to get the answers readymade but doing so may cause them not to use their imagination and cognitive faculty to solve the problem. There is always an impending need to finish the task before anyone else. It is good to have positive stress and healthy competition but the process has to be known correctly so that the right formula is applied and the dots filled to get the complete picture.

The success in filling these dots will depend upon the exposure of the student to similar solutions and knowledge of facts instead of just aiming for a high score. However, there has to be a perfect balance between the time taken to solve a problem and understanding the problem before writing the answer, so that the essence of the problem is not lost in the race to answer first. If there are two students, and one answers in 1 minute and does not explain how he derived the answer, whereas another student is able to do it in 2 minutes but is better at analysing the problem, the latter has a better chance to understand the problem as a whole. Therefore, rather than being just result oriented the tests should be process understanding oriented and grading or rewards should be appropriately calibrated.

This will not only ensure healthy competition but allow long term development of mental faculties which may assist in logical and rational thinking which is needed in higher classes and also in certain real life situations.

Question & Answer Series—III

Question No. 1:

What does it mean for students of mathematics to be persistent?

Response

Persistence plays a big role in learning difficult subjects sometimes and a student needs to be able to take failures and mistakes in his or her stride. Mistakes play a big role in helping us learn. For a student of mathematics, it is very important not to be afraid to try and even make mistakes, as it is only by persistence that one can turn adverse situations into success.

In fact, many personality development experts have stressed that persistence is one of the key factors that separate successful people from the ordinary. Many people tend to start a job enthusiastically but leave it halfway upon the first signs of failure. Then some other person picks up from where they left and tastes success. This shows that if one perseveres he is most likely to taste success sooner or later. So if the goal of a student is to learn a difficult subject or concept he must persevere with it.

Many of the greatest discoveries were achieved after trying repeatedly and failing repeatedly. So a student of mathematics must not give up if he fails initially, he must persevere and take the help of his teachers or seniors if required.

Question No. 2:

Is being quick always an indicator of intelligence?

Response

Being quick is not always an indicator of intelligence. An intelligent person finds ways of solving problems with minimal difficulty. It does not matter how fast or slow it is. Sometimes being fast in a situation can be detrimental to the targeted objective and being slow is just the right medicine for the malady.

Therefore, one cannot always measure intelligence in terms of how much time was required to solve a problem.

It may be true when you talk of computing power of a processor to be better if they are faster than slower ones while processing complex calculation. However it cannot be a predictor of the quality of the result obtained and the unquantifiable which may be a factor under consideration. If being quick leads to repeat processing for derivative or successive result it would be better to be slow to give a more conclusive result.

It has been proven that time is a relative component to the problem at hand; it may be wiser to be quick in certain case and slower in some other. Besides, there are various parameters to evaluate intelligence and quickness is just a part of it and not an absolute parameter. For example:

While driving a car at top speed may pay handsome returns on the racing circuit, it may give you speed tickets on the highway.

Participating in a quiz competition and answering quickly may be rewarding, but computing quickly on a production line and making errors may turn out to be costly if the product develops a fault due to it.

Therefore, being quick is not always wise and definitely not a sole indicator of human intelligence.

Question No. 3:
What ideas do you get about quality growth mindset for math tasks and to the saying Math is too much answer time and not enough learning time?

Response
The math task at hand can be calibrated differently and results may vary if there are more than one ways to solve the problem and everyone is allowed the freedom to choose any method that they wish to use. If we don't let creativity to flow it doesn't give enough to express and also it can bring in performance anxiety. So a healthy dose of creativity helps to answer problem at hand.

Growth Mindset Task Framework includes the following approached to learn.

1. Openness
2. Different ways of seeing
3. Multiple entry points
4. Multiple paths/strategies
5. Clear learning goals and opportunities for feedback.

Quality and growth mindset Math task pay more attention to the quality of analysing the scenario and arriving at a solution. This may allow the mind to take a different path and in the process give it an understanding of the process and allow a better understanding while arriving at a solution. If the class encourage this mode of learning it can be said quality or growth mindset approach to a problem at hand. Traditionally it is found that the focus is more on getting the answers right rather than learning as to how one derived a particular answer. This will provide initial success but will be detrimental in the long run. On the other hand growth mindset learning will allow quality learning irrespective of the time taken initially as will prepare better for future when one will be exposed to analysis and quality based questions and not answer based problems.

So Math may be too much of an answer time and not enough learning time if the class objective is purely result based.

Activity based questions.

Questions 4 and 5 are some activities which is open to the readers to answer. Here the growth oriented, quality based, open way and different approaches to the same problem at hand is encouraged. The idea is to visualize a formula based theoretical question and answer it with pictures and explanation. once a mind gets comfortable approaching the issue this way, we can then get a different dimension of approaching the same problem which we were not able to get in the usual rote and rigid method.

Question No. 4:

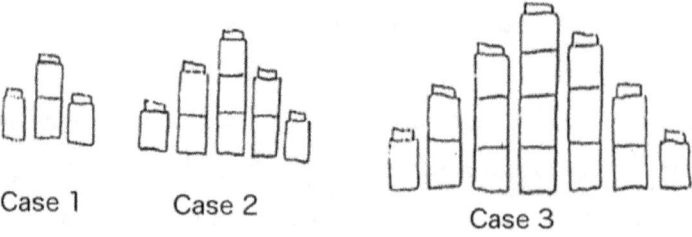

Case 1   Case 2   Case 3

A) How do you see this shape growing?

B) How many cubes are in case 100?

Response

A)

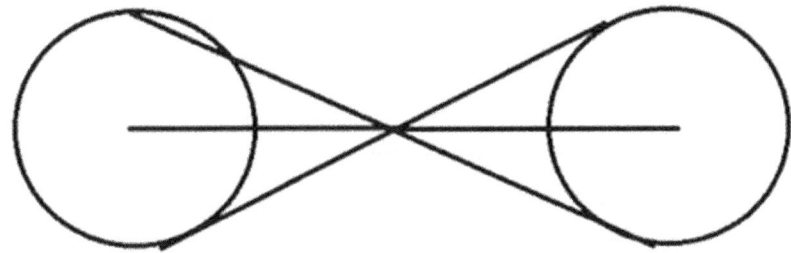

EXPLANATION (*In own words may differ or be incorrect, i have tried but without any proof is just an attempt, what i want to show how you have to approach a problem at hand irrespective of worrying about the answer and focus on learning to approach. Then with practice one may start getting positive results and answers. The veracity of the response is solely my own and has not to be taken as an authentic answer to the problem*)

A—Exponentially, pyramidcally, also the angle is moving from acuteness (i.e. less than 90 degrees) towards 180 degrees and further it can become obtuse (i.e. more than 180 degrees) and a complete circle i.e. 360 degrees and also beyond if matter permits and this is one way top to bottom and similarly bottom to top till the entire sphere or shape disintegrates to something else. This may happen in much advanced state. Like a cone with a circle down upside down to downside up. This can be done with any shape and number of cubes symmetrical or asymmetrically placed with shape of circle skewing i.e. the points on the circumference not being necessarily equidistant from the centre giving various oblong shapes of circles.

Answer B—101 x 1 + 100 x 2 + 99 x 2 + 98 x 2 + ... 1 x 2

Or

[(101 x 1) + 1)](+ 1 on top of middle most cube building and + 2 on cube symmetrically on either sides of the middlemost cube building is for

75

the small rectangles on the cubes which is displaying the same properties as a cube except in a cube all sides are equal and 90 degrees whereas in a rectangle all sides are not equal but the angles are all 90 degrees.)

So,

[[101 x1] + 1] + [[100 x 2] + 2] (100 multiplied by 2 as 100 cube buildings will be on either side of middlemost building to give proper symmetry and + 2 are small rectangle toppings on the either side cube buildings) + ... [[1 x 2] + 2]

B) How many cubes are in case 100?

Response

As interpreted by me

case 1 to 100	Total cubes
1	4
2	9
3	16
4	25
99	10000
100	10201
Formula n	$(n + 1)^2$

Where in formula n= number of cube replication case on both sides and the middle one is 1.middle one is number of cubes buildings to be replicated + 1 and squaring equates the number on both sides and gives a whole symmetry.

Look at the task framework and write about the dimensions that the two versions of the task ("How do you see the shapes growing?" and "How may cubes are in the 100th case?") differ on.

Response

The shapes growing task frame work talks about the results expected and want us to use the tools of laws to find variations and are open to interpretation and can be confusing if goal is not decided at the beginning. That's why it's more about asking the right questions than to get the right answers.

Whereas the question which specifically asks how many cubes are in the 100th case expects a specific answer based on the specific dimension.

It depends upon the questioners mind as to what he is expecting from you and if we are not on his wavelength we will get distorted version of answers. It does not tell the result expected by using the specific tools of mathematical law.

Therefore if the goal is performance as the desired expectation then the result of first question will be accepted even if results achieved are not specific. Whereas the aim of the second question is result oriented therefore the answer has to be specific irrespective of the steps performed.

For result based specific questions rules and formula should be known in order to achieve specific answers. However open ended question which are open to interpretation and have never been solved earlier will have to be inferred based on repeated observation before accepting it as a truth and then use it as a basis to deduce the problem further.

For example, if we are asked how does two parallel lines to grow if allowed to grow exponentially. Logic says that however big a line is it will not meet anytime. Now this is accepted as truth by repeated observation also called as a process of inference. Then making this a truth or an axiom or rule to fall upon other specific questions are asked.

However if we say that the two parallel lines meet at some point and grows to a cone, or a single line or any shape you have in mind you will not only have to assume certain facts by repeated observations with the help of other recognised or observed facts until you infer that to be true with proof and then base future problems based on that assumption.

This is the complexity which maths has.

Say for example you know earth is round cause it's proved by earlier observations and also shown from outer space, but if it is described as flat and proved by other facts to be flat will that make earths shape any different. The laws which existed will exist, for those who are closer to what it is accepted as by earlier proofs will stand better at getting specific results, whereas others have to invent the rules before proving the result to be any different.

Therefore true answer can be one or many.

Question No. 5:

Find an example of your own—a task that starts off as closed and short and encouraging fixed-mindset thinking to one that becomes a growth-mindset task.

Response

For example:

Analysis of number system as to why figures change when we add, multiply, subtracts and divides. More the former two, viz Addition and Multiplication.

If we add $1 + 1 = 2$ but $1 \times 1 = 1$, $2 + 2 = 4$ also $2 \times 2 = 4$, but $3 + 3 = 6$ but $3 \times 3 = 9$.

The closed and short mindset doesn't ask why they just follow it as it's expected to be that way. There are no arguments.

But growth mindset tries to find out visually what happens on addition and what on multiplication and why is squaring, square root, cube root (in advance math) give different answers. Why we need to work the calculation, when do we add and when do we multiply or square or divide or subtract. How do they get results differ marginally or greater. What is the block in numbers? How it invented and what are the drawbacks of a number system (decimal) these are the advanced questions which will emerge. But caution is one can get lost in them and lose track of the goal. So even growth mindset should know the limits, based on correct questions asked.

Sometimes explaining by example works better.

$3 + 3 = 6$ that is $(1 + 1 + 1) + (1 + 1 + 1) = 6$

But 3x3=9 that is we break three in to units (1 + 1 + 1) (group of friends) and then take another set of (1 + 1 + 1) (chocolates). 1ST PERSON in group wants all three chocolate. Then what will other two get nothing. So to ensure equitable distribution divide 1 chocolate in 3 equal parts total 9 parts for 3 chocolates. Now give 1 small part of each chocolate to each person, this will leave equal part for the other friends and ensure everyone gets to taste every chocolate. The quantity individually is less but each one will get equivalent to one full length of chocolate. This can be visually shown for better explanation also.

Question No. 6:

Think of some key things that children need to do to be able to learn mathematics easily. Write a few statements to show students what is important to know.

Response

The key thing is to have a map of learning before learning. It's very important to learn about the tools and system before one operates a machine. Similarly, one should be taught to navigate through the chapters and titles to understand the syllabus needed to be learnt. That's why children find navigating on video games and computers more stimulating and interactive than simply downloading lectures in a classroom scenario or reading books. The way video games function is it gives a great visual balance and rules on a page without being bothered to get correct result. As the player explores the game he fails but learns to win earlier than he even realizes and by the time he gets in to the grove he is already flying with system navigation quite sub-consciously. Similarly if learning maths can be as stimulating and without any fear psychosis it can be inculcated more easily using the sub-conscious and the conscious faculties of the mind.

However, there are proponents of reading as well. In fact, many accomplished computer programmers and researchers were book lovers. Reading helps one visualize the words and it may help if the concepts are clearly written down. Nowadays, surfing articles on the Internet or reading e-books also helps in a similar way.

The most important thing to know is what is important from an academic point of view and what is required for future research and development. If the goal is success in an examination, then a more focused approach is required. If acquiring knowledge is the goal, navigating through the wealth of information and finding satisfactory answers should be the motto.

Question No. 7:

What resources do you think may be valuable to you as a teacher or parent?

How can they support a growth mindset? (Explain with the help of the following question: If a store has to install a security camera, what angle will allow maximum surveillance?)

Response

I think the optimizing security camera question fits the problem at hand, talks about the practical application required, and then based on the syllabus taught in school, throws open the question to identify the best angle to watch over the customers. It tests the students on concepts such as line of sight, angle of vision, the percentage of area visible, and comparision of various positions. Thus it's an analytical question. It has to very clearly describe the concepts which it is expecting to be learnt before the analysis and its evaluation.

Thus the things to watch for are: 1) Concepts, 2) Tools required, 3) Resources, 4) Questions, 5) Answers given by students, 6) Expert evaluation, 7) Optimal solution, 8) Giving solutions as per concepts taught, 9) Give out-of-the-box solution, 10) Extra motivation for using out of syllabus concepts over and above the solutions and tools given11)Field of study that can be influenced by the problem at hand and solutions for it and, 12) Feedback given.

Thus some of the above resources that appeared will be valuable and the rest can be utilised for a growth mindset approach to study and learning.

Question No. 8:

Why do you think emphasis on getting positive results in exam, results in lower overall achievement for students in different studies?

Response

Emphasis on getting positive results in exam, results in lower overall achievement for students because the objective was to get academic results and not wholesome or overall development. There are schools where the student's all-round development is more important. Again, there are special schools for special needs; thus if the result expected is different from what is offered the achievement will also differ not necessarily lower or higher.For eg. If a school emphasize on merit percentage to achievements in sports to their final result of all students one will find the management and administration approach to skew more towards the former.This may not be bad but its just different from a school who emphasize sports in their overall grading of the student towards his final performance in school.

Similarly a school for special children or differently abled students will and must have a different calibration of overall student ratings to ensure pass results.

Also, even if one is put in a special group to move up the learning curve, it will result in reduced self-esteem whether one likes it or not. The fact that a child has been put in a different group because of his or her poor performance itself is a big de-motivator. Most of the time, the expected benefits are not achieved due to performance anxiety and also due to lack of motivation.

Being constantly tracked will not work unless there is a very good coach who will ensure that not only does the student feel motivated to be in the group and to perform better without being depressed about it. The student already knows he is weak in the subject and putting him on tracking reinforces the negative feeling unless he is coached effectively. Therefore, I feel it is best if the student does not know the reason for being tracked.

Question No. 9:

Design something different you will do with your students or your own children.

Response

Praise a child but not to the extent that he or she stops learning. Also, the main motivators are found to be cultural and in certain communities, being intelligent or smarter is their ticket to come out of poverty and help their families and their communities. This shows that the success achieved as per worldly standards was wired consciously and then due to sheer focus and hard work, the person became successful and rich. But it's not always that rich and successful people are the smartest. It never was because they just do things differently or have used their abilities better than others. It is that we assume that a successful person is more intelligent, which may not be the case from a mere academic point of view.

Success is a combination of various attributes and studies and mathematics is just a part of it. So if you want to produce a brilliant kid, there is no smartness pill for it is more about upbringing and the value system of a particular culture and the opportunities utilised and choices made by the student and the guardians. It's best to be assertive without forcing the child to behave in a particular way. Many of the brightest minds have found a way to channelize their energy and ideas to their and society's best advantages to become more successful. Teaching a child to be more humble, compassionate and emotionally intelligent can be a better matrix than solely relying on intelligent quotient as an indicator of one's abilities.

So I will try to design a programme to ensure that a child accepts reality, learns to be humble and knows that it is okay to fail, but it is not okay to be complacent, it is okay to be competitive but it is not okay to be caught up in the race, is energetic enough to move forward and is not lazy and cynical, and is an assertive person and not an argumentative idiot.

Question No. 10:

What do you think of "number-sense"?

Response

Most major concepts in mathematics are based on basic fundamentals. If there is "number-sense" and the manner in which mathematics is taught is correct, then lots of things that come later on can be beautifully included in the concepts. There are gaps in any new learning if we do not know how it fits into the bigger picture. However, if we are on the same wavelength as the teacher or author while reading a book, then it is fun.

Formulas are meant to give a universal language if we know the keys used in them correctly. If my key is different from yours and if I don't know what a particular sign represents, then it becomes difficult to interpret. Generally, different schools of thought and professors of a particular institute have a particular language and interpretations of the concepts; they may use an alpha, beta, or some other Greek variable or sign or symbol, but it is the underlying principle that is important. Most theorems and corollaries are once written in words and once in symbols. A theorem gets accepted if it uses the universally accepted language and is proven by laws of equation or mathematics. If a particular theory is proven wrong, then a new theory has to include better ideas or has to get rid of the problem in the previous idea.

In engineering or mechanics, it is a known fact that no design or machine is foolproof. There is always room for improvement and creativity. The error may be reduced from 0.1 to 0.000001, but as far as the numbers exist and the universally accepted rules apply, there is room for learning and creativity and new inventions and discovery. There are some groups who even try to change the rule book or change them as is evident from the use of new number systems in electronics and computer science which rely on binary number system rather than decimal system and there are other number systems as well. So the best learning appears if your fundamentals are right and you have done your research or have received the best guidance to learn what is right.

Number-sense allows this basic understanding upon which all advanced concepts are eventually based. One who knows this can flourish on the path to learning mathematics.

Question No. 11:

Give three reasons why students should discuss mathematics with others?

Response

Discussing mathematics with others help get various ideas for solving the same problems. It widens the student's mental horizons and it also saves time. Mathematics is used intuitively in our day-to-day chores. It's subtle, yet it's interesting but people make it sound hard and formal. Many of its applications are not formal at all.

The big idea I heard is that math can be used the way you understand it. One should have a growth mindset which never stops learning and is ready to dissociate itself established facts and norms. But that can cause a lot of trouble and that's why it needs courage to have a growth mindset. Very few people have the courage to fail and grow from it and still survive, because they tend to lose interest and initiative as they feel it's not worth the difficulties faced. Older ways of doing things are great but we do not remember everything. So there is always a scope to learn and unlearn from each other.

The big takeaway from the discourse is that a curious mind itself is a very big asset and makes the difference between being mediocre and great. This curiosity can be satisfied by taking initiative or by taking help from outside. It definitely helps if the student has the patience to go through the process of analysing and solving that question. Some may be lucky to have resources available at hand while others may not, but a curious mind will itself initiate that process of learning and gain from it irrespective of the availability of resources.

Therefore discussing with others will help a student get a different perspective, encourage teamwork and collaborative skills, and help develop healthy competition, all of which will help them in real life.

Question No. 12:

What kind of relationship do you want your students/children to develop with mathematics?

Response

The student should have a balanced view, subjectivity and objectivity and a curiosity to know more. They should have knowledge in the true sense, where they should be able to apply their knowledge and still feel thrilled about it. The motivation to learn should be there, but if a student is not into it, he should not be forced as he is just a child and needs other outlets. However, basic fundamentals should be learnt to get better at making decisions later on in life.

Labelling mathematics as a tough and cold subject while calling humanities a soft one, ignores the humanistic side of mathematics. So it's important to integrate math more holistically. One may not have the same goals, expectations, liking or conditioning, but that should not stop one from interacting with others who are different. It sometimes makes sense to let go than to cling to the problem and then get back with a fresh perspective.

Initial hiccups or delays in learning the basics will be more than compensated by exponential return in the long run. It will be intuitively adapted. Getting it tested by mathematical parameters and existing laws can then be assimilated once the base or foundation is correct.

Question No. 13:

Why should all students learn algebra?

Response

1. Meaning of Algebra-The part of mathematics in which letters and other symbols are used to represent numbers and quantities in formulae and equations.
2. A system of this based on given axioms.

All students should learn algebra to know the basics of the subject. It should not be compulsory if the field of study doesn't require it. However, if one intends to use it in future applications then it should be touched upon to give an understanding as to what it is. The difficulty is there are so many terms used even in definitions, which do not make sense unless explained with examples and applications.

It's not necessary to go into details but the picture should be very clear as to its usage and applicability. Also, the field has expanded and several sub-classes and systems have evolved thus diluting the purity of the subject.

At the same time, all students need not learn algebra. If there is an option available, one can opt out of learning algebra most algorithms have been computerised by various software programmes making the need to learn various equations redundant. However, if a student wishes to explore a field of study where the subject is essential, algebra should be learnt.

Question No. 14:

Tools to be used to describe a mathematical problem situation.

Response

1) As per the calculus textbooks developed by the faculty at Harvard, the Harvard Calculus consortium used the principle of 4 wherein, every situation should be described in at least four ways. From words, a picture, a graph, a table, and symbols.

2) The students can color-code the different representations.

For example, if you have an expression x + 3,

You could say, well, this is my plus 3 in the line, in the graph. This is the plus three in the visual of the cases. This is the plus three in the expression. So everything's color coded. And that's really helpful in teaching important connections. Connections and different representations lead to deeper learning and a focus on the structural, not only the procedural role of algebra. Even a multiple representation of one form, the symbolic form, can unleash incredible understanding.

Question No. 15:

How is growth quality approach to answering a mathematical problem different from the traditional methods especially when it comes to out of syllabus competitive(out of syllabus) exam questions? How does student who have the growth oriented approach training fair than the usual ones?

Response

It's due to the difference in approach in learning mathematics. In the top-to-bottom (quality growth) approach, students practise solving big problems and use their mathematical creativity. This system teaches them to visualize problems, and develop number-sense which leads them to develop a confident approach towards solving national level (competitive) exams which try to gauge these applicative skills in a student.

The other school which went the traditional way just make the students practise the methods but do not expose them to questions and real-life scenarios or the approach based way. This gives them confidence to only do things in the limited way they learnt. The basic concept to view the bigger picture is not practised and hence the less then favourable results.

Question No. 16:

How does solving mathematical puzzles help?

Response

Solving puzzles lead to visualizing the different ways of seeing the problems. It gives us the same brain exercise which summer workshops do for students. Concepts like generalizations, the curiosity oriented mind, the growth mindset, seeing a pattern in a problem, connections with mathematical sums, using intuitions—all these are used in games and puzzles.

The implications are pretty simple. It will empower a student to develop number-sense which will help him whenever the decision-making process is involved. It will make him confident and committed to undergo the process of building things and working on situations. Also, the open minded approach develops courage as mistakes are tolerated. The communication skills of the student will go a long way in covering up for things they don't know and will make them much sought after individuals.

Future

It is rightly said that the future of a thing does not only depends upon its past but also its present. In fact the time period is all connected from the past to the present to the future. We have seen how mathematics has taken shape right from the pre-historic and historic developments to the modern day. We have also discussed the present or contemporary issues at hand. Learning about the past helps us to know how things have developed. Though past trends are never a guarantee for future happenings, it gives us certain direction to look forward.

We have found old concepts re-visited and changed giving rise to newer theories and methods. There is a vast growth seen and various fields and subfields evolved from them. There are newer discoveries being made even today. The advent of digital age along with the power of computing has led to solving of complex equations which were thought to be impossible to solve until now. Educationists, administrators, schools, colleges and universities will have to reframe their curriculum and change their systems to include the applications relative to future growth. Many research organizations will help shape the future growth direction with the support of the various stake-holders.

We will find newer methods of teaching and learning evolve which will include the growth mind-set approach along with various tools with the help of modern day technology. All those areas which were visited earlier may be re-visited and history may repeat itself and come to a full circle. Or it may radically move us towards a new direction. We may find any of these paths taken or it may all be parallel and then depending upon the demand of the time, one of these approaches will be followed.

We will find many mathematical topics combining with existing areas of study and create a hybrid area of study. Multidisciplinary evolution will be possible as the computing power will enable vast storage, exchange, and retrieval of data at very high speed. In fact we are already living in the future and it is evolving even as we are reading this book.

Conclusion

I would like to conclude this book with the hope that I am able to convey the meaning and importance of this subject to each and every person affected by it. Knowing Mathematics correctly will not only expel those well-known fears but also enable it to be friendlier that what it has now become. Most people including myself have a fear and awe about the subject.

The purpose of this book is to enable an overall view about the subject and arrive at a comfortable attitude towards the subject. All those students and adults who hitherto hated the subject will now be able to appreciate it, keeping in mind the overall effect it has on everybody.

Even though it seems to be abstract, there is a lot of history and art associated with it. Initially most areas of study were restricted to the philosophies or sciences and then later on when it was specialised, it was classified separately into new fields of study. The way a particular theory or method is accepted depended on the political clout and reputation enjoyed by a particular professor or philosopher. And if somebody used a different method and made a discovery or innovation it was applauded. In fact the more applications one is able to clamour from a particular Mathematical principle or theory with the help of proofs, the more successful it is considered. If a particular discovery leads to invention of a technology or method which provides the highest common good, then that particular discovery emerges as the most successful field of study.

The problem is that students and academicians who do not directly control the flow of industry gets caught in it and does not know which path to choose and follow. Whether they should learn the already known things or should they question it and learn the other version. To get to this understanding requires a guide who can describe how everything falls in to place to connect the dots and get a bigger picture. Once you know where you stand you can get the map of things and set your destinations aligned to your goals. Once this is achieved learning become less burdensome and more fun.

It is this spirit of learning and self-discovery that I wish I could instil in the minds of people who go through this book.

If I have succeeded in getting things a little clearer I would have achieved what I had set out to do.

God Bless You All . . . . :-)

References

Most of the sources for this book are given below. If the reader wants to acquire more knowledge, he can visit the following websites:

1) http://en.wikipedia.org/wiki/History_of_mathematics

2) https://en.wikipedia.org/wiki/Babylonian_mathematics for your reference.

3) http://en.wikipedia.org/wiki/Maya_civilization

4) https://www.livescience.com/2427-amazing-aztecs-math-whizzes.html

5) https://math.temple.edu/~zit/Native%20American/9%20Aztecs_num.pdf

6) https://news.sciencemag.org/paleontology/2008/04/how-aztecs-did-math

7) http://en.wikipedia.org/wiki/History_of_science_and_technology_in_Africa

8) http://www-history.mcs.st-and.ac.uk/HistTopics/Egyptian_numerals.html

9) http://en.wikipedia.org/wiki/Egyptian_mathematics

10) http://en.wikipedia.org/wiki/History_of_mathematics

11) http://en.wikipedia.org/wiki/Chinese_mathematics

12) http://en.wikipedia.org/wiki/History_of_mathematics#Medieval_European_mathematics

13) http://en.wikipedia.org/wiki/History_of_ mathematics#Renaissance_mathematics

14) http://en.wikipedia.org/wiki/Lists_of_mathematics_topics

15) https://class.stanford.edu/dashboard

16) https://class.stanford.edu/courses/Education/EDUC115N/How_ to_Learn_Math/info

About the Author

Dhaval Vyas, M.B.A (Finance), B.A. (Economics & Sociology), is a management graduate from ICFAI (Institute of Chartered Financial Analysts of India). He has more than ten years of experience working in areas such as market research and survey, sales, administration, purchase, production, customer service industry and real estate.

Currently he is pursuing a company secretary course. An avid reader, his hobbies include various drawing, painting, watching movies, listening to music, social networking besides taking interest in socio-political issues and being a social investor. He has participated for a charity to remove poverty called Rangde which makes investments in the social sectors. He has an analytical bent of mind and has completed an internship for a project on the cement industry in India.

Vyas is also a keen conservationist and loves to learn about different cultures. This is his first book.